ANSWERMAN

Principles of Air Conditioning

Published by:

esco institute

AF808781

This manual was developed by The ESCO INSTITUTE, Mount Prospect, IL 60056

ESCO Institute
P.O. Box 521
Mount Prospect, IL 60056

Phone: (800) 726-9696 **Fax:** (800) 546-3726

Website: www.escogroup.org

E-Mail: customerservice@escogroup.org

Printed in the United States of America

ISBN Print Edition: 1-930044-03-8

Contents

Formulas

Temperature & Pressure

What is Temperature?

Temperature is an indication of the speed at which molecules move in a substance. It is ***not*** the measurement of heat, but rather the intensity of the heat. The phenomena of hot and cold is fundamental to humans since the first fire or ice appeared. Finding ways to measure temperature was a challenge for many great minds. It is a relative term based on water freezing at 32°F (0°C) and boiling at 212°F (100°C). These are fixed reference points on a thermometer (a device for measuring temperature). (See Figure 1.)

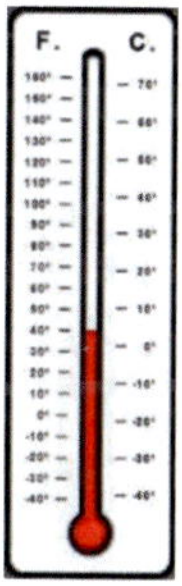

Figure 1

Range and accuracy have always been an issue when measuring temperature. The lowest possible temperature is minus 460°F (minus 273°C), called absolute zero. At absolute zero, all molecular motion ceases and all heat has been removed. Temperature tells us how hot something is but not how much heat it contains. For example, if a one-gallon container of water and a ten-gallon container of water are placed over identical heat sources, the temperature in the one-gallon container will rise faster than the temperature in the ten-gallon container. Ten times the heat energy must be added to the larger container in order to heat the water in both containers to an equal temperature.

Some may think that infinity is the opposite of absolute zero, but it is really "Planck Temperature". As of today, the highest possible temperature at which matter can exist is approximately (1.416785×10^{32}) Kelvin. Planck temperature absolute scale uses natural units with "0" as absolute zero and "1" as the hottest possible temperature. Beyond this temperature everything turns to energy, all subatomic particles collapse and become individual black holes obliterating space and time.

How is Temperature Measured?

Temperature is measured with a thermometer. Thermometers may be graduated in one of four different scales: Fahrenheit and Rankine used in the United States, Celsius (Centigrade) and Kelvin are used in the metric system. On the Fahrenheit scale, water freezes at 32°F and boils at 212°F. On the Celsius scale water freezes at 0°C and boils at 100°C.

For scientific calculations, absolute temperature scales without negative numbers must be used. On the Rankine scale, absolute zero is 0°R. The freezing point of water is calculated as 492°R (32°F + 460) and the boiling point of water is calculated as 672°R (212°F + 460). On the Kelvin scale, absolute zero is 0°K. The freezing point of water is 273°K and the boiling point of water is calculated as 373°K (100°C + 273). The Kelvin scale is also called Celsius Absolute (CA).

What is Ambient Temperature?

Ambient temperature is the measured temperature of the air surrounding an object. The components of a residential split air conditioning system are located both inside and outside the home. The measured temperature of the air inside the structure would be considered the inside ambient, and the measured temperature of the air outside the structure would be considered the outside ambient.

What is Dry Bulb and Wet Bulb Temperature?

Dry bulb temperature is actual air temperature and is measured with a dry bulb (ordinary) thermometer. Wet bulb temperature considers the enthalpy and the amount of moisture in the air. It is measured with a special thermometer that has a wick wetted with distilled water. Evaporation of the distilled water causes the wet-bulb thermometer to give a cooler reading than a dry-bulb thermometer. Less moisture in the air means that more evaporation occurs from the wick and therefore more cooling occurs; this results in a lower temperature reading. The difference between the dry-bulb and wet-bulb temperatures of the ambient air is called the wet-bulb depression. As the air's moisture content increases, there is less evaporation from the wick and less wet-bulb depression. When there is no wet-bulb depression, the relative humidity is 100 percent. Wet bulb temperature can also be measured with an electronic sensor and meter.

What is Dew Point Temperature?

The temperature at which condensation forms is called the dew point temperature. Air can be dehumidified by passing it over a surface with a temperature below the dew point, such as an evaporator coil. Moisture collects on the cold surface of the evaporator coil and is drained through the condensate line.

What is Pressure?

Pressure is defined as force per unit area and in the United States is most commonly described as pounds per square inch. The SI metric system uses a pressure unit called the pascal, named for French scientist Blaise Pascal. A pascal is equal to one newton per square meter (N/m^2). A newton is the force required to cause a mass of one kilogram to accelerate at a rate of one meter per second.

The operation of an air conditioning or refrigeration system depends mainly upon pressure differences throughout the system. The physical properties of a substance are affected by its surrounding pressure, especially its boiling point. Air conditioning technicians must be able to understand and work with pressures above and below atmospheric pressure.

What is Pascal's Law?

Pascal's Law states that pressure applied on a confined fluid will be transmitted equally in all directions. This is the basis of operation of all hydraulic and pneumatic systems. When air is pressured into a balloon it expands in all directions.

What is Atmospheric Pressure?

Earth's atmosphere extends about fifty miles above the planet's surface and consists of approximately 78% nitrogen and 21% oxygen. The remaining 1% is composed of other gases. Even though gas molecules are very small, they still have weight. The atmosphere exerts a pressure of 14.7 pounds per square inch at sea level. At higher altitudes, atmospheric pressure is significantly less. The most common way of measuring atmospheric pressure is with a mercury barometer. Normal atmospheric pressure at sea level (14.7 psia) will support a column of mercury 29.92 inches high. At standard atmospheric conditions the weight of air is 0.075 lbs. per 1 cubic foot.

What is Gauge Pressure?

The pressure measurement we commonly use is called gauge pressure. Atmospheric pressure on the gauge is indicated as 0 psi or psig (pounds per square inch gauge). Compound gauges used to measure low-side pressures in air conditioning systems can measure pressures both above and below 0 psig. Gauge readings are relative to atmospheric pressure. Compound gauge readings below atmospheric pressure are indicated in inches of mercury (Hg). It may be necessary to adjust compound gauges periodically to compensate for changes in atmospheric pressure. These types of gauges have an accuracy of 1% to 2% of the full scale. A gauge that measures from 30Hg to 800 psig with 1% accuracy can be off by as much as 8.15 psi.

What is Vacuum?

Any pressure below atmospheric is considered a partial vacuum. A perfect vacuum is the absence of all atmospheric pressure. An in-depth understanding of vacuum principles is necessary for the technician to perform proper refrigeration equipment service and installation. Pressures below atmospheric are usually measured in inches of mercury (in. Hg) or millimeters of mercury (mm Hg). An increase in pressure increases the boiling point of a liquid, and the opposite is also true. A lower pressure results in a lower boiling point. An air conditioning or refrigeration system is evacuated to a deep vacuum to vaporize and remove moisture from the system. A micron gauge is used for measuring deep vacuums. A micron is 1/1000 of a millimeter of mercury and one inch of mercury is equal to 25,400 microns.

What is Absolute Pressure?

With the absolute pressure scale there are no negative numbers, the scale starts at 0 pounds per square inch (psi). It is necessary in field service and in engineering calculations to use the Ideal Gas Law. The absolute pressure scale allows the measurement of both vacuum and pressure to be made using the same units. Absolute pressure measurements are indicated as pounds per square inch absolute (psia). 0 psia is the lowest possible absolute pressure. Because atmospheric pressure is 14.7 psia at sea level, gauge pressure can be converted to absolute pressure by adding 14.7 (15 rounded off) to the gauge pressure measurement. Suction and discharge operating pressure must be converted to absolute pressures to calculate compression ratio.

Basic Thermodynamic Principles

What is Refrigeration?

Refrigeration is the movement of heat from an area where it is not wanted to an area where it is less objectionable. For example, a refrigerator transfers heat from the inside of the refrigerated cabinet to the kitchen's ambient air. A heat pump is a system used to move heat in two directions. For cooling, heat is transferred from inside air to outside air. For heating, heat is transferred from the cold outside air into the home's inside air.

What are the Laws of Thermodynamics?

The laws of thermodynamics give us a basic understanding of heat. The first law of thermodynamics states that energy cannot be created or destroyed; it can only be converted from one form to another. This law applies to heat and to all forms of energy: electrical, mechanical, solar, chemical, and nuclear.

The second law of thermodynamics states that heat always travels from hot to cold. The speed at which heat travels from one object or substance to another depends upon the temperature difference between the two. A greater temperature difference causes a faster heat transfer. The transfer of heat will stop when both objects or substances reach the same temperature.

How does Heat Travel?

Since refrigeration involves the exchange of heat, it is important to understand how heat travels. There are three ways in which heat travels, conduction, convection, and radiation.

What is Conduction?

Conduction is the transfer of heat either through an object or by direct contact between two objects. When one end of a metal bar is heated, the heat travels by conduction from molecule to molecule to the cooler end of the bar.

What is Convection?

Convection is the transfer of heat through flowing medium in either a liquid or vapor state. Convection creates currents as the warmer (expanding) fluid rises above the colder (contracting) fluid. Convection currents can be natural or forced. Natural convection occurs as warm air naturally rises above cooler air. Forced convection is formed using a fan or blower to move air across a heat exchanger.

What is Radiation?

Radiation is light waves and radio waves. The heat from the waves are absorbed by solid objects, not by the space they travel through. Radiant heat from the sun travels through space and then is absorbed by the earth and objects on the earth. For example, the heat produced inside a car or being sunburned is from the sun's radiant energy.

Physical States

What are the Three States of Matter?

Matter can exist as a solid, liquid, or gas, depending on pressure, temperature, and heat content. In all substances above absolute zero, molecules are constantly in motion. Even when matter is in its solid state, its molecules are moving very slow. In the liquid state, molecules move faster. If liquid is heated to its boiling point its molecules move so fast that they are now free to travel in all directions creating a gas.

Does Heat Effect Change of State?

As a substance is heated or cooled, it may undergo a change of state. The substance most commonly used to illustrate this is water. At atmospheric pressure with temperatures below 32°F, water exists in its solid form, ice. Between 32°F and 212°F water is a liquid, and at 212°F it can exist in a liquid or vapor state. With the addition of heat, it becomes steam.

Saturation point is the term used to describe any substance at its boiling point at a given pressure. Material in a saturated condition contains as much heat as it can hold without changing state. Atmospheric pressure has a direct influence on the saturation point (boiling point) of liquids. For example, water at sea level boils at 212°F; however, on a 15,000-foot-high mountaintop, it boils at 184°F. For another example, placing a radiator cap on your automobile raises the saturation point of the water approximately 2.5°F for each pound of pressure increase.

Since different materials have different molecular structures, the temperature at which a change of state occurs varies. For example, the saturation point of water is 212°F at atmospheric pressure, but the saturation point of refrigerant HFC 410A is -57°F at atmospheric pressure. If the pressure is controlled, the saturation temperature can be controlled. This is done in a refrigeration system to create the operating temperatures of ultra-low, low, medium, and high.

What is Btu?

Because a standard of measurement for quantity of heat was necessary, the international standard British Thermal Unit (Btu) was created. The Btu is a measurement of the exact amount of heat necessary to raise the temperature of one pound of water one-degree Fahrenheit. This is also the base quantity of heat used to compare the amount of heat it takes to change the temperature of other materials, and this quantity is known as specific heat (sp.). The weight of a material or substance, specific heat, and the amount of temperature change is used for calculating total Btu.

Example:
15 lbs. of water changed from 70°F to 90°F.
Btu = Lbs. x sp. x temperature change (ΔT)

Btu = 15 x 1 x (90-70)
Btu = 15 x 1 x 20
Btu = 300

The Btu per hour is used to rate refrigeration and heating equipment. 12,000 Btu per hour equals one ton of cooling. 1 ton of ice melting at 32 degrees absorbs 12,000 Btu per hour.

The metric system uses the calorie and gram. The calorie is a measurement of the exact amount of heat necessary to raise the temperature of one gram of water one-degree Celsius.

Sensible and Latent Heat

What is Sensible Heat?

Sensible heat is heat energy that causes an increase or a decrease in measured temperature. Its name comes from the fact that it can be sensed as a change in temperature. A change in sensible heat can be felt with your hand or measured with a thermometer.

A thermometer measures only the temperature of a substance. It does not measure the amount of heat required to reach a certain temperature.

What is Latent Heat?

Latent heat is heat added to or removed from a substance causing a change of state without a change in temperature. For example, water heated to 212°F may be either liquid or gas. To change liquid to gas, more heat must be added. Remember, the temperature of water remains at 212°F as it boils. Additional heat is needed to change the water molecules from liquid to steam. Latent heat is sometimes called hidden heat because it cannot be felt by hand or measured with a thermometer. There are five types of latent heat:

- latent heat of melting
- latent heat of fusion
- latent heat of vaporization
- latent heat of condensation
- latent heat of sublimation

Every substance has a particular temperature and/or pressure point at which it changes state from solid to liquid or from liquid to vapor. At 212°F, 970 Btu are required to change one pound of water into steam. To change the steam back into liquid, 970 Btu must be removed.

This same principle applies to latent heat of fusion. At 32°F, 144 Btu must be removed in order to freeze one pound of water. It takes the same amount of energy (144 Btu per pound) to melt one pound of ice. Latent heat of vaporization and condensation are the basic principles behind most refrigeration systems. As refrigerants absorb heat, they change from liquid to gas in the evaporator. The heat is then removed from the gas, changing it back into liquid in the condenser.

Some substances change directly from solid to gas with no visible evidence of changing into a liquid state. This phenomenon is latent heat of sublimation. For example, dry ice (composed of carbon dioxide, or CO_2) is one such material.

What is Superheated Vapor?

A vapor that has been superheated is at a temperature above its saturation point (boiling point). When a substance is superheated, sensible heat has been added to the vapor. For example, if steam is heated to 213°F at atmospheric pressure, it is superheated 1°F.

The air around us consists mostly of nitrogen and oxygen. Both gases have boiling points well below 0°F at atmospheric pressure. Therefore, we can say that most of our atmosphere is made up of superheated vapors.

What is a Subcooled Substance?

Any substance whose temperature is below its saturation point is considered subcooled. Water that is 210°F at atmospheric pressure is subcooled 2°F. In other words, subcooling is the number of degrees a substance has been cooled below its boiling point.

Gas Laws

What are the Gas Laws?

A technician needs an in-depth knowledge of a gas's behavior in order to accurately understand and diagnose system problems. Air conditioning systems constantly change liquid to gas and gas to liquid as the method to move heat. There are three factors which affect the laws of gases: absolute temperature, absolute pressure, and volume. Gas laws demonstrate that a gas will always behave according to the rules, with no exceptions.

What is Boyle's Law?

Boyle's Law states that the pressure of a gas varies inversely to the volume, provided the temperature remains constant. For example, the cubes in Figure 2 show that the pressure will double if the volume is cut in half.

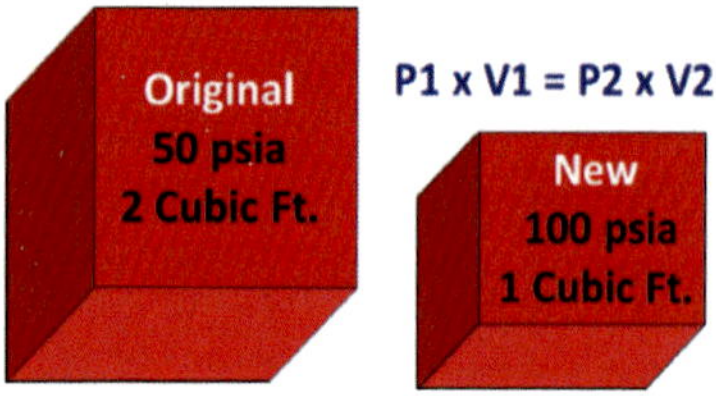

Figure 2

Original pressure multiplied by original volume must equal new pressure multiplied by new volume. For the original cube 50 psia multiplied by 2 cubic feet = 100, with the new cube changed to 1 cubic foot the pressure is equal to 100 psia or 100 multiplied by 1.

What is Charles' Law?

Charles' Law states that at a constant pressure, the volume of a confined gas varies directly to the absolute temperature. Also, at a constant volume, the pressure varies directly to the absolute temperature. When pressure remains constant, the volume will increase as temperature increases. If volume remains constant, pressure will increase as temperature increases as shown in Figure 3.

$$V1 \times T2 = V2 \times T1$$

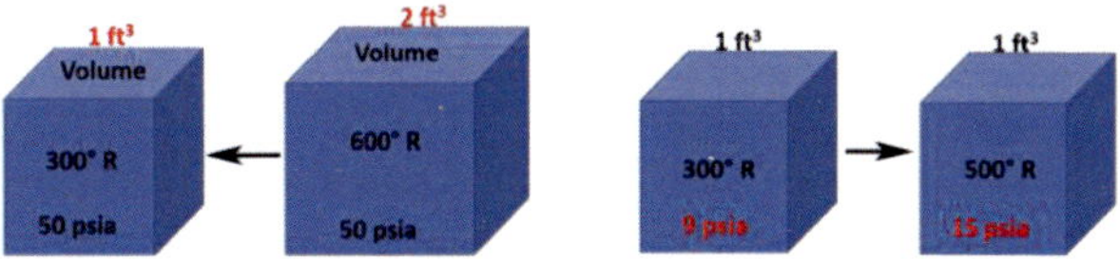

Figure 3

What is Dalton's Law?

Dalton's Law states that the total pressure of a confined mixture of gases is the sum of all the pressures of each gas in the mixture. The force of the pressure is exerted in all directions as shown in Figure 4.

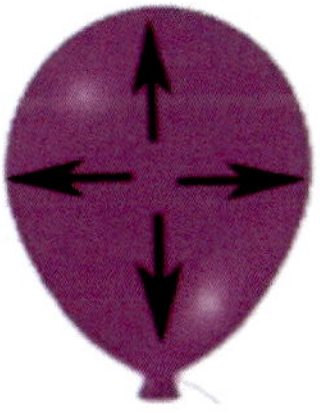

Figure 4

What is the Ideal Gas Law?

The ideal gas law is a combination of both Charles' and Boyle's Laws. Written as the product of the original absolute pressure times the volume, times the new absolute temperature, equals the product of new absolute pressure, times the volume times original absolute temperature.

Either of the following equations may be used.

$$P1 \; x \; V1 \; x \; T2 = P2 \; x \; V2 \; x \; T1$$

or

$$\frac{P1 \; x \; V1}{T1} = \frac{P2 \; x \; V2}{T2}$$

Note: Pressure and temperature must always be absolute.

Refrigeration Cycle and Components

What is the Vapor/Compression Refrigeration Cycle?

The compressor is the heart of the vapor-compression refrigeration cycle. Low-pressure, low-temperature, superheated refrigerant vapor entering the compressor is compressed. Compression changes the refrigerant to a high-pressure, high-temperature, superheated vapor. From the compressor it moves to the condenser where the heat is removed, de-superheating then condensing it into a liquid. The liquid refrigerant is subcooled to a point below the liquid saturation temperature before it leaves the condenser. The high-pressure subcooled liquid refrigerant flows to the metering device. As the refrigerant moves through the metering device, the liquid is reduced to a low-pressure causing a small percentage (approximately 20%) of the liquid to flash to a gas (flash-gas), which lowers the remaining liquid refrigerant to its saturation temperature. The low-pressure, low-temperature, liquid refrigerant flows into the evaporator. As the liquid refrigerant absorbs heat, it evaporates into a low-temperature vapor. During this process the refrigerant vapor at the end of the evaporator becomes superheated above its saturation temperature before entering the suction line. From the suction line, refrigerant enters the compressor as a low-pressure, low-temperature, superheated vapor to repeat the cycle. The compressor and the metering device are the dividing points between the low-pressure and high-pressure sides of the system. (See Figure 5.)

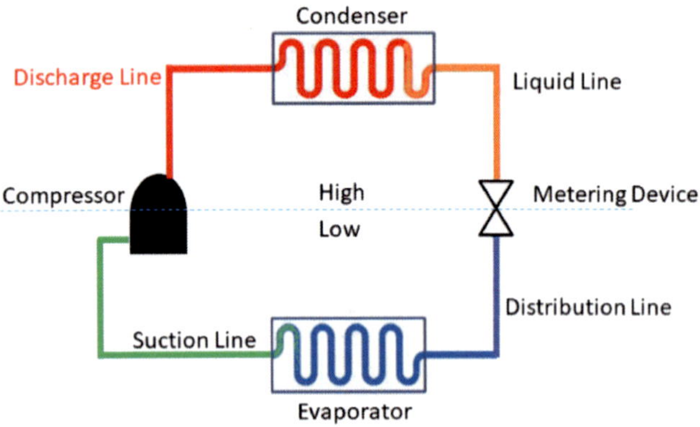

Figure 5

What is the Compressor?

As stated, the compressor is the heart of the vapor-compression refrigeration system. A compressor is a mechanical device that increases the pressure of a gas by reducing its volume. There are several different compressor designs, but each serves the same purpose, creating a pressure difference which becomes one of the dividing points in the system. A compressor performs two functions. The first is providing suction to maintain low pressure in the evaporator. The second is providing compression to convert low pressure/low temperature refrigerant from the evaporator to high pressure/high temperature vapor, or highly superheated vapor.

There are five types of compressors commonalty used in the refrigeration and air conditioning industry:

Reciprocating: Is a compressor with one or more pistons and cylinders. It is the most widely used type of HVACR compressor. These compressors can be open drive, semi-hermetic, and welded hermetic. They may be used in residential and commercial air conditioning and refrigeration.

Applications include household refrigerators and freezers, air conditioning, commercial and industrial refrigeration.

Rotary: A rotary-vane compressor consists of a cylindrical casing, one suction and one discharge opening with a rotor positioned eccentrically with respect to the casing. The multi-vane rotary is used in large refrigeration applications as a booster compressor.

Applications include appliances such as household refrigerators and freezers, residential air conditioning, and heat pump products usually below 5 tons.

Scroll: A scroll compressor is used to compress refrigerant to a high pressure and temperature using a fixed and orbital scroll meshed together.
Applications include residential and commercial air conditioning.

Screw: A screw compressor uses rotors which are comprised of the pair of meshing screws to compress large volumes of gaseous refrigerant to a high pressure and temperature.

Applications include commercial and industrial air conditioning and refrigeration.

Centrifugal: A centrifugal compressor or a radial compressor, compresses refrigerant by rotating impellers.

Applications include large commercial and industrial air conditioning and refrigeration systems.

What is Compression Ratio?

Compression ratio is calculated by dividing the absolute suction pressure (psia) into the absolute discharge pressure (psia). Compression ratio is an important factor when considering a refrigerant and temperature range for a particular application. A single-stage compressor for low temperature usually requires a compression ratio of 10 to 1 or lower. A high temperature system such as air conditioning requires a lower compression ratio of approximately 3.5 to 1. Very high compression ratios that exceed 10 to 1 ratio require the use of a two-stage compression. A two-stage compressor will have two compression stages, 1st cylinder with a large piston to the 2nd cylinder with a smaller piston.

discharge pressure psig + 15 psi = discharge psia
suction pressure psig + 15 psi = suction psia
discharge psia ÷ suction psia = compression ratio

Example:
418 psig + 15 psi = 433 psia discharge pressure.
125 psig + 15 psi = 140 psia suction pressure.
433 psia ÷ 140 psia = a 3:1 ratio.

A dirty evaporator can have a noticeable effect on compression ratio. Lower suction temperature and pressure causes the refrigerant vapor that the compressor is pumping to become less dense, lowering the compressor performance and increasing the compressor temperature, resulting in higher energy use.

Poor condenser efficiency such as insufficient air flow, blocked or restricted coils, will affect the high side pressure and the compression ratio. The increased head pressure causes the amount of refrigerant gas in the clearance volume of the compressor to be greater than design conditions allow. Increased discharge temperature accelerates the chemical reactions that form acids and will cause oil break down.

What is Inverter Capacity Control?

An inverter takes an alternating current (AC) power supply and coverts it to direct current (DC). Electronic switches in the inverter pulse the DC output varying the frequency to simulate a three-phase power supply. Varying the frequency to an AC motor or electronically commutated motor (ECM) controls the rotation and speed of the motor. Without an inverter, the AC motor would operate at full speed as soon as the power supply was turned on. Inverters electronically control the electrical voltage, current, and frequency of electrical devices such as the compressor motor in an air conditioner. Information is received from sensors which monitor operating conditions and adjust the speed of the compressor motor, which directly regulates air conditioner output. The use of an inverter to adjust the speed and acceleration of a motor increases the range of applications for the motor, compared to a motor that operates at a constant speed. The speed of a motor is measured as the number of revolutions per minute (rpm). The acceleration rate is given as the change in speed over a specific time period.

What is the Condenser?

A condenser is a heat exchanger used to remove heat from the refrigerant to another medium such as air or water. The condenser removes heat from the refrigerant vapor causing it to condense back into a liquid. The first function of a condenser is de-superheating the entering refrigerant vapor so that condensing can occur. As the refrigerant cools giving up heat, it condenses, becoming liquid. The final stage of refrigerant through the condenser causes the liquid refrigerant to become subcooled. The amount of subcooling is determined by the refrigerant charge and condenser design.

There are three basic condenser types: air-cooled, water-cooled, and evaporative.

Water-cooled condensers condense refrigerant at about 105˚F when the water is supplied at 85˚F with a leaving temperature of 95˚F, resulting in a 10˚F approach temperature.

Depending on the design efficiency, an air-cooled condenser condenses refrigerant at approximately 15˚F to 30˚F above entering air temperature. For example, an R-410A air-cooled condenser would have a head pressure of 390 to 445 psig on a 95˚F day depending on the efficiency of the condenser.

(95˚F ambient +15˚F = 110˚F) = 365 psig
(95˚F ambient +30˚F = 125˚F) = 447 psig

What Controls the Refrigerant?

Every refrigeration system needs some type of refrigerant metering device to control the flow of refrigerant. The refrigerant metering device and the compressor separate the high-pressure side of the system from the low-pressure side. High pressure/near ambient temperature liquid enters the inlet side of the metering device. The metering device causes a drop-in pressure, and a small percentage of the liquid refrigerant flashes to vapor causing the remaining liquid refrigerant to change to a low temperature saturated liquid. The metering device also maintains the correct amount of refrigerant in the evaporator. Too much refrigerant allows liquid to reach the compressor which can cause serious damage. Too little refrigerant prevents the system from working efficiently.

What is the Capillary Tube?

A capillary tube is the simplest refrigerant metering devices consisting of a copper tube with a small inside diameter. Capillary tubes have no moving parts and does not require adjusting once it is properly sized.

The length and diameter of the capillary tube determines the refrigerant flow and pressure drop based on the compressor capacity and type of refrigerant.

A capillary tube is a fixed metering device so the system cannot adjust quickly to load changes. The capillary tubes are best suited for use in applications where the load remains relatively constant. One advantage to using a fixed-bore capillary tube is that it allows system pressures to equalize during the off cycle. This reduces the starting torque requirements of the compressor motor.

Because the inside diameter of a capillary tube is very small, measured in thousands of an inch, equipment manufacturers usually place a strainer or drier in the line just before the tube to prevent debris from blocking refrigerant flow.

Capillary tube systems are classified as critically charged systems. Refrigerant charge is best analyzed when a unit is operating at design conditions. When a unit is operating with the proper amount of refrigerant, the superheat can range from approximately 5 to 30 degrees depending on the application. Some small appliance manufacturers take advantage of heat exchange by soldering the capillary tube along the length of the suction line. In this case, a superheat reading must be taken at the evaporator outlet before its point of contact with the capillary tube.

Capillary tube or fixed orifice systems are slow to respond to changes in load and charge adjustments. The system will take at least fifteen minutes to reach a balanced condition after a charge adjustment is made. For small systems manufacturers recommend weighing the correct refrigerant charge into an evacuated system.

What is a Fixed Orifice Piston?

A fixed orifice (piston actuators) is a very simple metering device. Its design is a piston with a fixed bore, seat, and retainer inside a body. Its size or flow rate must match the condensing unit capacity. Due to its ability to allow a counter flow of refrigerant makes it a viable metering device used on heat pumps. In the cooling mode the piston acts as a check valve and the refrigerant must flow through the fixed bore. In the heating cycle the flow is in the opposite direction and the refrigerant flows around the outside of the piston.

The charge for split system air conditioning or heat pump in the cooling mode is calculated by system superheat. An overcharged fixed orifice can increase operating pressures, lower evaporator superheat, decrease efficiency, and also lead to compressor flooding. Undercharged fixed orifices can lower pressure, increase superheat, lower efficiency, lead to excessive oil in the evaporator from lower refrigerant flow, and cause compressor overheating. The fixed orifice metering device will allow the high and low side pressures to equalize and does not stop the liquid flow into the evaporator during off cycle. This is the primary reason for having a critical charge.

What is a Thermostatic Expansion Valve?

A thermal expansion valve (TEV, TXV) is a mechanical metering device that controls the flow of refrigerant into the evaporator. A thermostatic expansion valve is one of the most commonly used metering devices in air conditioning and refrigeration applications. Thermostatic expansion valves make it possible to keep an evaporator operating under varying load conditions while maintaining the proper superheat. They can be used in a wide variety of applications. By maintaining a constant degree of superheated vapor leaving the evaporator, thermostatic expansion valves are often referred to as constant superheat.

Three main operating pressures govern the operation of a thermostatic expansion valve: spring pressure, evaporator pressure, and bulb pressure. Evaporator pressure and spring pressure push up on the valve diaphragm as closing forces. The bulb pressure is transmitted to the top of the valve diaphragm through a tube acting as the opening force.

The sensor bulb is attached in close thermal contact to the outlet of the evaporator on the suction line. The bulb senses the temperature of the suction vapor at the outlet of the evaporator at the point of bulb contact. Bulb pressure is transmitted to the top of the valve diaphragm through its tube. As the evaporator superheat increases, the corresponding increase in bulb temperature and pressure works against evaporator and spring pressures forcing the valve to open. The increased refrigerant flow lowers evaporator superheat to the set point. As soon as the temperature drops at the outlet of the evaporator the sensing bulb cools, and the pressure is reduced on the top of the TEVs diaphragm. The valve will modulate opening and closing to control the superheat setting.

Several valve and bulb charge designs have been developed to provide maximum valve performance for each application. Some TEV designs have a spring pressure that is factory set and non-adjustable, while others provide for field adjustment. When making superheat adjustments, technicians should allow approximately 10 to 15 minutes after each adjustment for the system to reach a balanced condition before another adjustment is made. A properly sized valve will maintain a very even pressure and superheat. An oversize valve will constantly open and close causing erratic pressure and superheat. The constant opening and closing of the valve is called hunting.

What is an Evaporator?

An evaporator is a heat exchange device located in the area where cooling is desired. Liquid refrigerant and some flash-gas is fed into the evaporator from the metering device at a reduced pressure. Once in the evaporator, the liquid refrigerant goes through a latent change of state by absorbing heat from the air or medium passing over the evaporator. The lower temperature of the evaporator causes the moisture in the air to cool to its dewpoint temperature condensing water vapor back to water.

Proper airflow and clean coils must be maintained in order to ensure adequate heat exchange and efficiency. Aluminum fins are used to increase surface area of the evaporator coils. The spacing is based on design temperature, the lower the temperature the farther apart the spacing. Air conditioning systems are designed so that all refrigerant vaporizes in the evaporator. Only superheated vapor should leave the evaporator. Nominal airflow is approximately 400 cubic feet per minute (cfm) per ton (12,000 Btu) for air conditioners to operate correctly. Dirty filters, evaporators, or improper fan operation can reduce the amount of heat absorbed by the refrigerant in the evaporator and may cause flooding of liquid refrigerant back to the compressor or increase compression ratio. Since liquids are essentially incompressible, liquid refrigerant allowed to reach the compressor can cause serious damage. During the heating cycle of a heat pump, the indoor coil is used as the condenser. Heat pumps require a higher volume of air. Air flow may be increased up to 450 cfm for a heat pump's indoor air handler.

What are Accessory System Components?

Accessory components are used to enhance system operation, improve efficiency, and keep systems clean and dry. Some of the most common components are suction accumulators, sight glasses, liquid line driers, suction line filter driers, and service valves. These devices are specific to the application and are used to serve or assist the general operating principles of the major components.

What is a Filter-Drier?

A liquid line filter drier is designed to absorb moisture, acid, and capture foreign matter through the use of screens and desiccants. Liquid line driers are designed to remain in the system and should be replaced any time the system is opened for service. These driers usually have soldered or flared connections. When liquid line driers ice up or begin sweating, this is a sign that they are clogged and must be replaced. Once the new drier is installed, the system must be evacuated to 500 microns prior to charging. On heat pump systems, a bi-flow type liquid line drier must be used. This drier type allows refrigerant to be filtered in both heating and cooling modes of operation. (See Figure 6.)

Figure 6

There are several locations where manufacturers install liquid line filter driers and direction of flow must be observed. Driers are usually marked with an arrow to indicate direction of flow.

Suction line driers are used to remove acid and sludge from the system after a compressor burnout. This gives the needed protection for the new compressor because the refrigerant is filtered right before it enters the new compressor. To prevent a repeat failure of the new compressor, it is important that contaminants in the system be removed. The suction line drier is ***not*** designed to remain in the system like a liquid line drier. Depending on the severity of the burnout the suction drier may require more than one replacement. After all the acid and sludge has been removed, the suction line drier should be removed from the system. Suction line driers have one or two gauge-ports to measure the pressure drop across the drier. As they begin to clog-up, the pressure drop will increase, and they should be replaced.

Refer to manufacturers literature to determine the proper size, type, use, and installation of all line filter driers. (See Figure 7.)

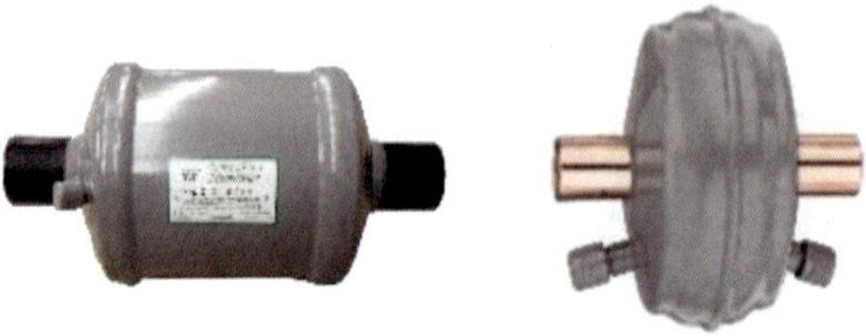

Figure 7

What is a Liquid Receiver?

The liquid receiver is installed between the condenser and the liquid line. It serves as a storage tank for liquid refrigerant. A liquid receiver is usually found on commercial systems that use a thermostatic expansion valve. (See Figure 8.)

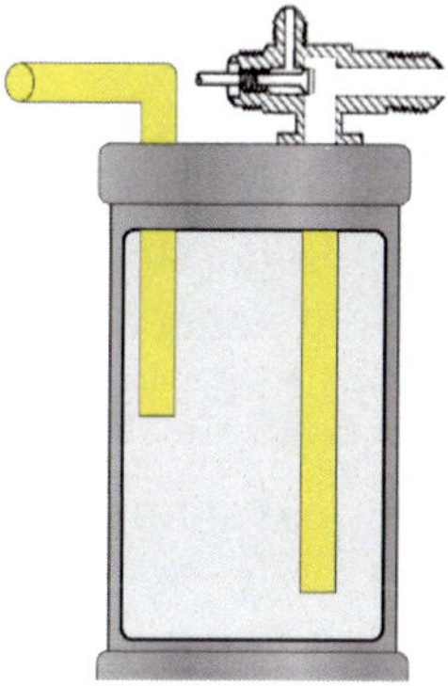

Figure 8

The refrigerant level in the receiver will change with the cooling load on the system. The receiver outlet is connected to a tube that extends down to the bottom of the receiver tank. This tube ensures that only liquid refrigerant is delivered to the thermostatic expansion valve at the various loads of the system. The receiver should be sized to hold approximately 15% or more than the systems full refrigerant charge. Most receivers will have a king valve to isolate the refrigerant in the receiver. Receivers can come in many sizes and are designed to accommodate the total refrigerant charge while also allowing space for expansion.

What is an Accumulator?

A suction line accumulator prevents compressor damage by preventing liquid refrigerant from entering the compressor from the suction line. The suction line accumulator is a tank designed to boil off liquid refrigerant and return it back to the compressor at an acceptable rate. Systems that use a fixed metering device have no way of controlling exactly how much refrigerant is sent to the evaporator. If all the refrigerant is not vaporized in the evaporator, liquid refrigerant could make its way to the compressor. Therefore, an accumulator is installed in the suction line just before the compressor to catch any liquid refrigerant and allows only vapor to reach the compressor. This may also apply to TEV systems or low-side float systems.

What is a Sight Glass?

The sight glass is a device for viewing refrigerant flow in the liquid line. The sight glass allows the technician to view the condition of the flowing liquid. Bubbles may be an indication of liquid refrigerant flashing to a gas; possible problems may be low refrigerant charge or a restriction in the liquid line ahead of the sight glass, or possibly a clogged filter/drier. Most sight glasses also have a moisture indicator centered in the viewing glass. The moisture indicator will change color when excessive moisture is present. The color change is dependent on the manufacturer. Some sight glasses must be replaced once moisture has changed the color of the indicator; while others may return to the color indicating dry once the system is evacuated. Sight glasses should not be used to evaluate the system charge when servicing the system. These devices are not an accurate method of evaluating charge. For example, an over charged system would have a clear sight glass, and a system fully charged with a blended refrigerant may have bubbles.

What are Service Valves?

Service valves are used to gain access to a refrigeration system. They may be of the stem type or Schrader valve type. The stem type valves are called king valve and two-way service valves. The king valve is normally closed and only opened for system service. (See Figure 9.)

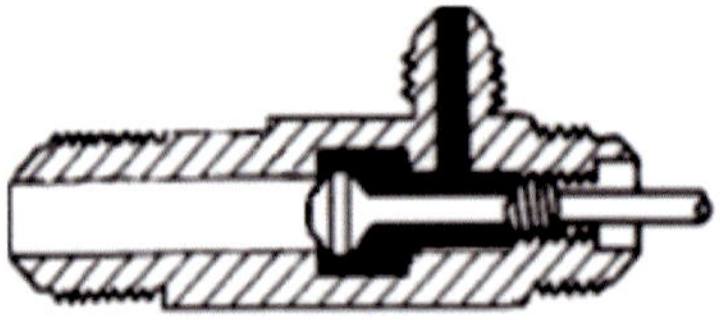

Figure 9

On the stem type two-way service valve there are three positions, back seated, cracked, and front seated. Notice the black section of the valve is what is open in the back-seated position. (See Figure 10.) The line port and compressor port are open, and the service port is closed. The black solid color is where the refrigerant will flow. This valve is in position for normal operation.

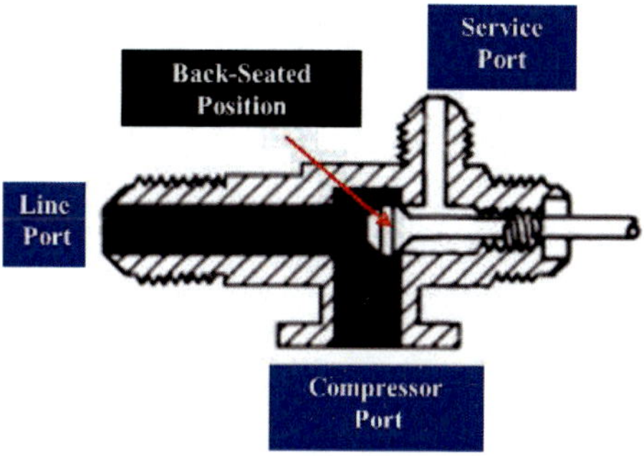

Figure 10

In the front-seated position, the black solid color is where the refrigerant will flow. (See Figure 11.) The line port is closed, and compressor port is open to the service port.

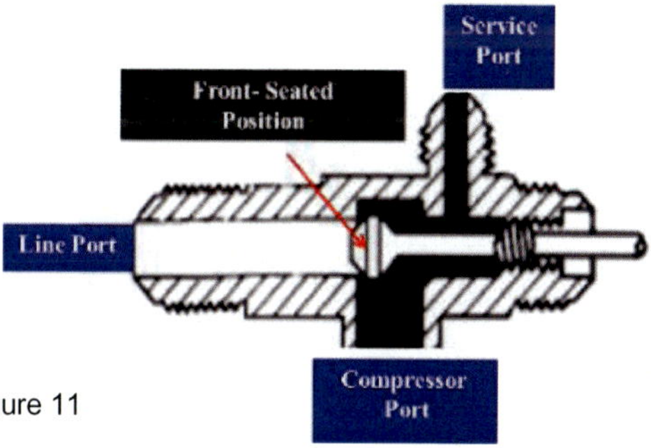

Figure 11

In the crack-seat position, all three connections are open. The line port, compressor port, and service port are all open to each other. (See Figure 12.) This position is for servicing and checking pressures.

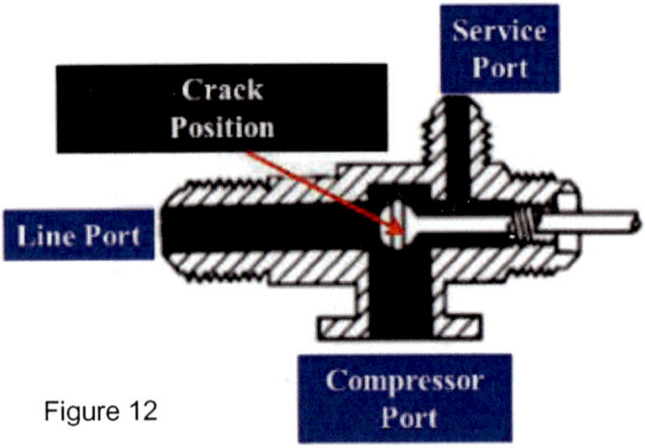

Figure 12

The two-position service valve is used by most air conditioning manufacturers. (See Figure 13.) This two-position valve has a Schrader access valve for the service port which allows access to the system without opening or

Figure 13

closing the main valve. This valve is normally used on residential condensing units. The valve is front seated when installed on the condensing unit at the factory to hold in the refrigerant charge.

The Schrader access valve is used on small systems that do not come with service valves. They are soldered in place after all the refrigerant is removed from the system allowing future access and the recharging of the system. (See Figures 14, 15.) Common sizes of the copper tube 1/8", 3/16", and 1/4".

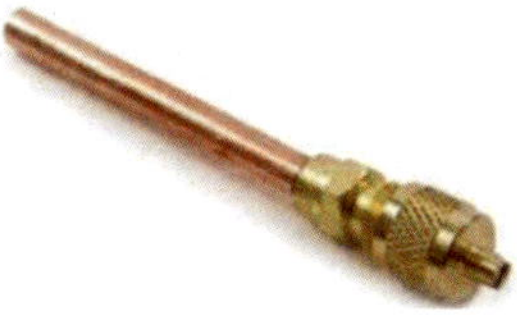

Figure 14

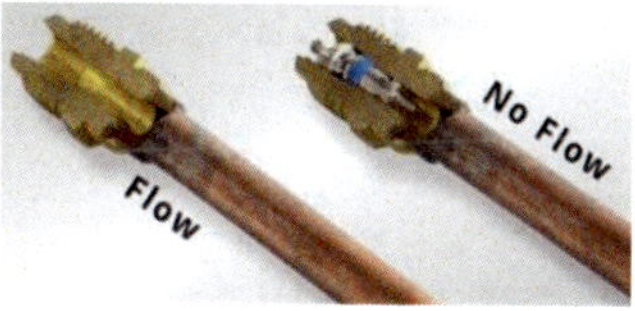

Figure 15

Piercing type valves are used for gaining temporary access to a sealed system. (See Figure 16.) They must not be left on any system and should be replaced with a Schrader valve. When installing an access fitting onto a sealed system, the fitting should be leak tested before proceeding with recovery. It is generally recommended that solderless piercing type valves only be used on copper or aluminum tubing material. Common sizes for copper tubing are 3/16", 1/4", 3/8", 1/2", and 5/8".

Figure 16

Refrigerants

What are Chlorofluorocarbons (CFCs)?

CFCs were the first of a new class of chemical refrigerants developed in the late 1920's and were used in a variety of applications. They consisted of Chlorine, Fluorine, and Carbon atoms. Chlorofluorocarbons are fully halogenated hydrocarbons that are colorless, nonflammable, chemically stable compounds.

The most common types of CFCs were dichlorodifluoromethane (R-12 or Freon-12) and trichlorofluoromethane (R-11). No one knew at the time that these new CFC compounds would lead to the destruction of the ozone layer in the stratosphere. The manufacture of such compounds has been phased out under the Montreal Protocol. In 1994, R-12, R11, and many other CFCs were phased out across the United States and many other countries. The chlorine in CFCs will neither dissolve in water nor break down into compounds that dissolve in water, so it does not rain out of the atmosphere. Naturally occurring chlorine will dissolve in water and rain out of the lower atmosphere. Chlorine released from CFCs destroys ozone in a catalytic devastating reaction where 100,000 molecules of ozone can be destroyed per chlorine atom. The chlorine atom, (Cl), behaves very differently from the chlorine molecule (Cl_2). The radical Cl. is long-lived in the stratosphere, where it attacks the ozone (O_3) changing it into O_2 and chlorine monoxide. Ozone absorbs UV-B radiation, so its depletion allows more of the sun's high energy radiation to reach the Earth's surface. Most CFC's can be decomposed by high temperatures (open flames, glowing metal surfaces, etc.) forming hydrochloric and hydrofluoric acids and possibly phosgene gas.

What are Hydrochlorofluorocarbons (HCFCs)?

HCFCs consist of Hydrogen, Chlorine, Fluorine and Carbon. In the 1950s R-22 came into the picture and began replacing R-12 in widespread usage for home and commercial air conditioning. R-22 operated at higher pressures than R-12, and in the beginning older technicians were skeptical and concerned about working with a higher-pressure refrigerant. By the early 1960s this refrigerant eventually became the main refrigerant used in home and commercial air conditioning for the next 50 years. HCFCs were less damaging then their counterpart CFCs but are also responsible for depleting the Ozone layer in the stratosphere and are being phased out.

The approximate phase-out dates for HCFCs are:

- Freeze from beginning of 1996
- 35% reduction by 2004
- 75% reduction by 2010
- 90% reduction by 2015
- Total phase out by 2020

What are Hydrofluorocarbons (HFCs)?

Hydrofluorocarbons (HFCs) are the fourth generation of refrigerants that are fluorine based. These refrigerants contain Hydrogen, Florine, and Carbon chemicals. Many HFCs are mixtures that can be zeotropic, near azeotropic, or single-component refrigerants such as R-134a and R-32. R-134a has 2 hydrogen atoms, 2 carbon atoms, and 4 fluorine atoms, making it a pragmatic formula of tetrafluoroethene. The "a" suffix indicates that the isomer is unbalanced by one atom. R410A is a near-azeotropic blend of R-32 and R-125. The US Environmental Protection Agency recognizes it as an acceptable substitute for R-22 in household and light commercial air conditioning systems. HFC refrigerants are the zero-ozone depletion potential refrigerants, because there are no Chlorine chemicals in the compound. Ozone depletion or ODS is the destruction of the ozone layer that filters out harmful UV light. HFC refrigerants are a type of synthetic greenhouse gas, have medium to high GWP and contribute to global warming if released to the atmosphere. These gases reflect the solar heat that radiates off the Earth's surface back to the surface, instead of going back out to space, thus increasing the Earth's mean temperature. The two most common HFCs are R-134a and 410A. HFCs are used in a wide variety of cooling systems, from refrigerators and freezers to automotive air-conditioning units. The atmospheric lifetime of HFC-134a is approximately 14 years; CFCs, by comparison, can remain in the atmosphere for 100 years. R-32 is an HFC refrigerant that has an ODP of 0 and a GWP of 675 over the 100-year period by which refrigerants are rated. It has a flammability rating per ISO Standard 817 and is listed as a 2L refrigerant in ASHRAE Standard 34. However, it has a relatively low burning velocity that is slightly lower than ammonia's burning velocity, therefore it's not explosive, like butane or pure hydrogen.

What are Hydrofluoroolefins (HFO)?

Hydrofluoroolefins (HFOs) much like traditional hydrofluorocarbons (HFCs), are composed from hydrogen, fluorine, and carbon. The only difference is that they are unsaturated, meaning that they have at least one double bond molecule. These molecules are named olefins or alkenes, unlike HFCs these new refrigerants have very low effect on global warming. Olefins have carbon atoms connected by a double bond where alkanes have single bonds between carbon atoms. Hydrofluorocarbon (HFC) refrigerants like R-134a, are all comprised of hydrogen, fluorine, and carbon atoms connected by a single bond. Hydrofluoroolefins are a new class of refrigerants that have a much lower global warming potential than its HFC alternatives. One example, HFO-1234YF (See Figure 17.) an alternative to 134a, is 335 times lower on the global warming potential scale and only four times higher than carbon dioxide. HFOs are a new variety of synthetic refrigerants that are a promising solution to the environmental issues. HFO blends are not drop-in refrigerants, therefore changes must usually be made to the equipment before they can be used. HFOs are either A1 or A2L refrigerants, meaning they are either non-flammable or mildly flammable. HFO refrigerants have better miscibility properties with POE oil than did the HFC refrigerants. They are not miscible in alkylbenzene or mineral-oil lubricants.

R-Number	Chemical Name	Boiling Point	Cylinder Color
1234yf (CF3CF=CH2)	Tetrafluoro propene	-22°F	White/Red

Figure 17

What are Hydrocarbons (HC)?

Organic compounds are considered as refrigerants under this group. Most of them possess satisfactory thermodynamic-properties but are highly flammable. HC refrigerants are classified by ASHRAE in the A3 safety group, which are highly flammable. Pure HC refrigerants have no chlorine or fluorine in their molecules. They only contain hydrogen and carbon, having a zero-ozone depletion potential (ODP). HC refrigerants are highly efficient reducing energy consumption compared to HFC refrigerants. These refrigerants are often referred to as natural refrigerants. Many of the CFC, HCFC, and HFC refrigerants were formed using molecules of ethane and methane. R-600a (isobutane) a miscellaneous organic compound, R-290 (propane), and R-441A (HC blend) have been approved by the EPA's SNAP program for use in the U.S. Uses for these three HC refrigerants include new stand-alone retail refrigerators, new domestic refrigerators and freezers, or combination refrigerator/freezers. They can use isobutane (R-600a) or R-441A if the refrigerant charge size does not exceed 2.0 ounces. Freezer equipment can use propane (R-290) if the refrigerant charge does not exceed 5.3 ounces. (See Figure 18.)

R-Number	Chemical Name	Boiling Point
R-170 (C2 H6)	Ethane	-127°F
R-290 (C3 H8)	Propane	-44°F
R-1150 (C2 H4)	Ethylene	-155°F
R-1270 (C3 H6)	Propylene	-54°F
R-600 (C4 H10)	Butane	31°F
R-600a (C4 H10)	Isobutane	11°F

Figure 18

What is Carbon Dioxide?

Carbon dioxide (CO_2) (R-744) is a clear gas composed of one atom of carbon (C) and two atoms of oxygen (O). It is a stable, inert, and non-toxic gas. Carbon dioxide is one of many molecules in which carbon is commonly found on the Earth. It does not burn at standard temperature and pressure.

Carbon dioxide (R-744) has several appealing characteristics as a refrigerant. It is non-flammable, no ozone depletion, available in large quantities, and low cost. However, its downfall is low efficiency and a high operating pressure (approximately 10 times higher than R134a). For these reasons, efforts are needed to improve its refrigeration cycle technology, particularly in heat exchangers and expansion devices. Heat pumps could also benefit from CO_2 due to the higher temperature that can be obtained even at very low ambient temperatures. Plants take in CO_2, break it down into carbon and oxygen, release the oxygen to the atmosphere, and then retain the carbon to live and grow. This is very important because it is a key part of the Earth's carbon lifecycle.

What is Ammonia (R-717)?

Ammonia is a chemical consisting of one atom of nitrogen and three atoms of hydrogen. It is designated in a chemical notation as NH_3 or R-717 as a refrigerant. (See Figure 19.)

Figure 19

Ammonia is extremely soluble in water and is frequently used as a water solution called aqua ammonia. Anhydrous ammonia is basically 99.99 percent pure and is used as a refrigerant. (“Anhydrous" is a Greek word meaning without water). It is a highly irritating gas with a pungent, suffocating odor. Ammonia boils at minus 28°F and has latent heat of vaporization of 565 Btu per Lb. at 5°F. One cubic foot of liquid Ammonia at 60°F will expand to 850 cubic feet of gas.

Only Ammonia has stood the test of time, though very ***deadly,*** it is still the best refrigerant pound for pound used in industrial refrigeration with proper handling and strict safety rules. Ammonia has been used in industrial applications since the late 1800s and is generally acknowledged as being one of the most efficient refrigerants.

R-Number	Chemical Name	Boiling Point	Cylinder Color
R-717 (NH_3)	Ammonia	-28°F	Silver
R-744 (CO_2)	Carbon Dioxide	-109°F	Green
R-764 (SO_2)	Sulfur Dioxide	14°F	

Environmental Impacts

What is Ozone Depletion?

Stratospheric ozone depletion is a global issue that can lead to problems such as crop loss, skin cancer, increased eye diseases such as cataracts, and reduced plankton and other microscopic marine life.

CFCs (chlorine, fluorine & carbon) and HCFCs, (hydrogen, chlorine, fluorine & carbon), which contain chlorine, have been found in air samples taken from the stratosphere. When CFCs and HCFCs, are released into the atmosphere their **chlorine** content causes depletion of the ozone layer. When a chlorine atom encounters an ozone molecule, it takes one oxygen atom from the ozone forming a compound called chlorine monoxide (ClO), leaving an oxygen O_2 molecule behind. The chlorine monoxide will collide with another ozone molecule releasing its oxygen atom, forming two O_2 molecules, leaving the chlorine free to attack another ozone molecule. A single chlorine atom can destroy up to **100,000 ozone molecules**.

What is Global Warming?

Another problem brought on by man's abuse of nature is a phenomenon known as the "greenhouse" effect. This is largely the result of carbon dioxide (CO_2) and other gases released from burning fossil fuels. These gases hold heat in the atmosphere in much the same way as heat is held in the interior of a car parked in the sun with the windows rolled up. Release of gases from refrigeration systems, industrial pollution, automobile exhaust emissions, paints, cleaning solvents, etc., and the destruction of trees and plant life which are beneficial to the atmosphere, has added to the buildup of undesirable gases. The resultant global warming would have catastrophic results.

For all these reasons, the EPA requires that the refrigerant be recovered using proper equipment before a system is opened for service. Used refrigerants can be stored in an approved container (recovered), cleaned for reuse by oil separation and filter/driers (recycled), or processed to a level equal to new refrigerant specifications (reclaimed).

A fine of $44,539 per day per violation may be enacted for breaking the laws of the Clean Air Act.

What are the Refrigerant Recovery Rates?

REQUIRED LEVELS OF EVACUATION FOR TYPE II & III APPLIANCES
[Except for small appliances, MVACs, and MVAC-like appliances]

Type of appliance	Using recovery and/or recycling equipment manufactured or imported	
	Before November 15, 1993	After November 15, 1993
Very high-pressure appliance	0" Hg	0" Hg
High-pressure appliance, or isolated component of such appliance, with a full charge of less than 200 lbs. of refrigerant.	0" Hg	0" Hg
High-pressure appliance, or isolated component of such appliance, with a full charge of 200 lbs. or more of refrigerant.	4" Hg	10" Hg
Medium-pressure appliance, or isolated component of such appliance, with a full charge of less than 200 lbs. of refrigerant.	4" Hg	10" Hg
Medium-pressure appliance, or isolated component of such appliance, with a full charge of 200 lbs. or more of refrigerant.	4" Hg	15" Hg
Low-pressure appliance	25 mm Hg absolute	25 mm Hg absolute

System Performance

What is a Gauge Manifold?

One of the most important tools for an HVACR technician is the gauge manifold set. The left side, compound gauge (blue) and the right side, high-pressure gauge (red), are attached to the manifold to measure system pressures. Hoses are used to connect the manifold to the refrigeration system's access ports or service valves to gain access to system pressures. The compound gauge measures low-pressure (psig) and vacuum (inches Hg). The high-pressure gauge measures high side (discharge) pressure. Depending on the refrigerant, the high-pressure gauge may be rated at 500 or 800 psig. The manifold is also equipped with a center port, (usually a yellow hose), that can be connected to a recovery device, evacuation vacuum pump, or charging device.

What is a Digital Gauge Manifold?

Digital manifold gauges can display all the readings a technician needs at once. It is a wireless, 4-port manifold, and micron gauge for HVACR professionals.

This digital gauge measures 2 pressures, 2 temperatures, and vacuum in microns. (See Figure 20.)

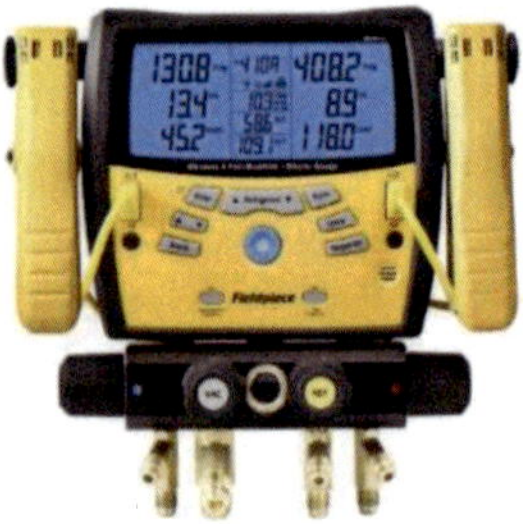

Figure 20

It allows the technician to view live pressure, target superheat, and automatically calculate real time superheat and subcooling. This allows the technician to see how the charge is affecting the system in real-time for system diagnosis and charging. It has a built-in micron vacuum gauge which gives technicians the ability to view the micron rate of change. Some digital gauges have a mini-usb port for updating to newer versions of firmware. This allows the technician to update in the field and ensures the most up-to-date features for the manifold. Be sure to purge your manifold and hoses before connecting to a system with a different refrigerant. Always follow manufacturer's directions when cleaning vacuum and pressure sensors.

EPA recommends that hoses be equipped with low loss fittings or valves that manually close or which close automatically to minimize refrigerant loss when hoses are disconnected. The hoses used for service and with recovery equipment must be equipped with low-loss fittings.

How do I perform Leak Testing?

Leak testing is best performed prior to evacuating and charging the system. A visual inspection of the entire system may reveal traces of oil at fittings and connections. Open drive compressors are prone to leakage at the drive shaft seal.

Some technicians rely on a standing vacuum test to find a leaking system. This is not only time consuming, but it is also inefficient because of the low-pressure differential between the vacuum and atmospheric pressure (approx. 15 psi.). Pulling a vacuum on a system with a leak will result in contaminating the system with moisture and air.

In years past, refrigerant was used to leak test the system. This has proven to be damaging to the environment as well as a waste of valuable refrigerant. The most efficient method of leak testing a system is to use compressed dry nitrogen. Nitrogen is an inert gas that will not harm the atmosphere.

Never pressurize a system with oxygen or compressed air, the oxygen mixed with refrigerant oil can cause an explosion.

Always pressurize nitrogen through a pressure regulator.

Compressed nitrogen cylinders are under very high pressure, approximately 2400 psi in a full cylinder. Use a pressure regulator to reduce the pressure to the manufacturer's recommended test pressure. Leak testing with nitrogen presents a new problem. Electronic leak detectors are sensitive to the halide elements found in most refrigerants. These detectors will not detect leaks of pure nitrogen; therefore, it is necessary to add a small amount of refrigerant as a "tracer" gas.

Another alternative is to use an ***ultrasonic leak detector***. These leak detectors have a highly sensitive headset to ***"listen"*** for leaks as well as LED indicators.

Make sure the ***electronic leak detector*** is compatible with the type of refrigerant in the system and that it is properly calibrated. Electronic leak detectors designed for halogenated hydrocarbon refrigerants can ignite a flammable refrigerant and cause an explosion.

When testing for a leak, the leak detector probe should be moved at approximately 1 inch per second. Moving the probe too quickly can lead to missing a leak.

Use of a ***bubble solution*** is a third alternative that is safe to use on most systems. There are soap solutions that are specially made for leak detection that will indicate a leak, whereas house hold soaps will not.

Whatever method is used, ensure that the system is leak free before evacuating and charging with refrigerant.

Low-pressure systems use open type centrifugal compressors and operate in a vacuum; leaks in the gaskets or fittings will cause air and moisture to enter the system. Leaks must be repaired to reduce the operation of the purge unit and loss of refrigerant. The most efficient method of leak checking a charged, low-pressure, refrigeration unit is to pressurize the system using controlled hot water or heater blankets. When controlled hot water or heater blankets are not feasible, use nitrogen to increase system pressure. When pressurizing a system, do not exceed 10 psig. Exceeding 10 psig can cause the rupture disc to fail. When leak testing a water box, be certain the water has been removed before placing the leak detector probe through the drain valve. To leak test a tube, use a hydrostatic tube test kit. Systems with open drive compressors are prone to leaks at the shaft seal.

What Are the Leak Repair Requirements?

Section 608 Leak Repair Regulations

The updated Section 608 regulations include new leak inspection and verification test requirements for owners/operators that will affect technicians starting January 1, 2019:

Leak inspections are required for appliances that have exceeded the applicable leak rate, according to the schedule below. All visible and accessible components of an appliance must be inspected, using a method or methods that are appropriate for that appliance.

Equipment	Full Charge	Frequency of Leak Inspections
Commercial Refrigeration and Industrial Process Refrigeration	Containing greater than 500 pounds	Once every three months until the owner/operator can demonstrate through leak rate calculations that the leak rate has not exceeded 20% (commercial refrigeration) or 30% (IPR) for four quarters in a row.
	Containing 50 to 500 pounds	Once per calendar year until the owner/operator can demonstrate through the leak rate calculations that the leak rate has not exceeded 20% (commercial refrigeration) or 30% (IPR) for one year.
Comfort Cooling	Containing50 or more pounds	Once per calendar year until the owner/operator can demonstrate through the leak rate calculations that the leak rate has not exceeded 10% for one year.

Initial and follow-up verification tests: leak repairs are required for appliances that exceed the applicable leak rate. The verification tests must demonstrate that leaks were successfully repaired.

- An initial verification test must be performed before any additional refrigerant is added to the appliance.
- A follow-up verification test must be performed only after the appliance has returned to normal operating characteristics and conditions. There is no minimum timeframe.

What is System Evacuation?

Evacuation is the removal of water vapor and non-condensables from the refrigeration system with the use of a vacuum pump. The vacuum pump removes moisture by reducing pressure so that the moisture will vaporize at normal ambient temperatures.

In the HVACR field the term "Vacuum" is used to express the amount of pressure inside a refrigeration system when preparing for refrigerant charging. A perfect vacuum is the removal of all (100%) atmospheric pressure. This is not possible with the vacuum pumps used in HVACR industry. The excepted vacuum level is 500 microns or lower in the HVACR field. When 500 microns or lower is achieved, the system is considered evacuated (dry). It is extremely important to understand that non-digital refrigeration gauges do not measure vacuums accurately or precisely.

A thorough understanding of vacuum and pressure principles is an absolute necessity for the technician. Since an increase in pressure will increase the boiling point of a liquid, the opposite is also true. Lower pressure will result in a lower boiling point. (See Figure 21.)

Figure 21

What are System Evacuation Terms?

Non-condensables: Gases that cannot condense in the condenser.

Evacuation: The removal of water vapor and non-condensables from the system.

Dehydrate: The removal of water and water vapor from the system.

Vacuum: Refers to the absence of air and water vapor in refrigerant system.

Micron: A micron is a unit of measurement used for measuring a vacuum.

25,400 microns equal 1Hg

Two inches of mercury equal 1psia

The amount of moisture a system will tolerate is determined by the type of system, type of refrigerant, and the type of oil. Under all circumstances, moisture in a refrigeration system must be kept to an absolute minimum. Moisture is the number one factor in acid formation, sludge build up, oil breakdown, and corrosion. Moisture can freeze in the refrigerant control valve and block the flow of refrigerant or hold the control valve open and flood the evaporator. Most refrigeration system failures can be traced back to moisture problems.

Moisture will react in different ways depending on the material it comes in contact with. Three of the biggest reactions that occur within an air conditioning or refrigeration system are the reactions between metal (polished steel, copper, or aluminum), moisture, refrigerant, and oil.

Metal + Moisture = Corrosion

Moisture + Heat + Refrigerant = (hydrochloric and hydrofluoric acids)

Oil + Refrigerant + Acid = Sludge

Cleaning after burnout: contaminants form within the system from the mixture of moisture, acid, and oil. This buildup results in sludge. Sludge cannot be removed from the system using a vacuum pump. Standard cleanup procedures using filter-driers should be used.

Deep vacuum: refers to pressures below 29 inches of mercury or 1 psia. Low pressure is normally measured with an electronic micron gauge. Single and two-stage vacuum pumps are commonly used in the HVACR field today. The two-stage model pulls a deeper vacuum because the first-stage exhaust is vented into the second-stage intake then out of the pump. This allows second stage to pull lower pressure, increasing effectiveness and producing a lower vacuum. Vacuum pumps are rated in cubic feet per minute, at the largest fitting on the pump. Normal requirements are a 1 Cfm pump for every two tons of refrigeration making two to eight Cfm pumps the most commonly used.

What is the Evacuation Procedure?

After recovering refrigerant from the system or installation of a new system, connect the center hose from the gauge manifold to the vacuum pump and the low and high side hoses to the respective service valve. It is recommended that if a system has a Schrader valve the core be removed using a valve core removal tool.

Note: It is extremely important that a system containing (A2L) refrigerants has had the refrigerant charge completely removed prior to beginning any evacuation procedures with a vacuum pump.

Check all connections before starting the vacuum pump. Slowly crack open both manifold hand valves. The system pressure should pull down to 500 microns. If the pump does not pull adequate vacuum, the system may have a leak or contaminated oil saturated with moisture. The preferred method of measuring a deep vacuum is to use an electronic micron gauge. A compound gauge is not accurate at deep vacuum levels.

The vacuum pump should be allowed to run for at least one-half hour at 500 microns to allow enough time for any moisture to vaporize. Large systems may require more time to achieve the required vacuum. There is no such thing as over evacuation of a system.

Before shutting off the vacuum pump, close both manifold valves, then shut off the pump. The system should now hold a vacuum. The system pressure should not rise above 1000 microns, or manufacturer's recommended rise, during the decay test.

A ***decay test*** is a specified period of time that the vacuum is monitored for a rise in pressure.

What is the Deep Vacuum Procedure?

Refers to pulling a system down to as low as 500 microns. This process of pulling a vacuum to 500 microns is accomplished with the use of a two-stage vacuum pump, and large diameter vacuum hoses. (3/8 or 1/2-inch diameter) (See Figure 22.)

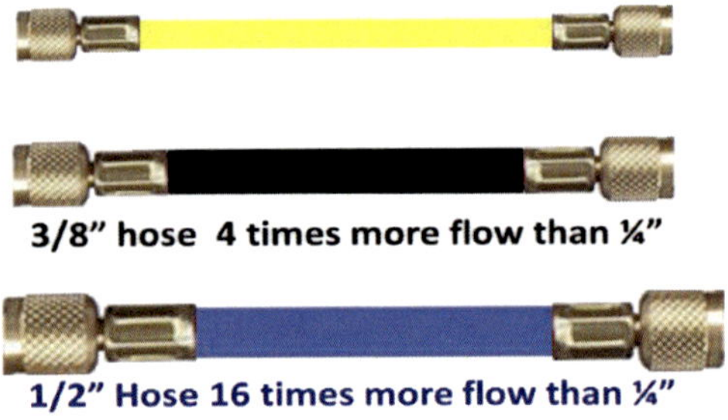

Figure 22

1. Where possible purge the system with nitrogen or inert gas. This is done by purging through the liquid line port and venting off at the suction port.

2. Make sure the system pressure is at zero. Connect vacuum pump to the manifold and system service valves. The micron gauge should be connected as far away from the vacuum pump as possible.

3. Start the vacuum pump.

4. Once the system is evacuated, the vacuum pump is isolated. If the system pressure begins to rise according to the micron gauge, there may be a leak, trapped refrigerant, or moisture still in the system.

What is the Triple Evacuation Procedure?

1. Purge the system with nitrogen or inert gas. This is done by purging through the liquid line port and venting off at the suction port.

2. With all nitrogen purged and pressure at zero, connect vacuum pump to the manifold and system service valves. The micron gauge should be connected as far away from the vacuum pump as possible. Evacuate to 1500 microns.

3. Break the vacuum by introducing nitrogen or inert gas to a positive pressure (maximum 15 psig) and allow 5 to 10 minutes for nitrogen to absorb moisture.

4. Purge the nitrogen or inert gas from the system. Reconnect the vacuum pump and pull a vacuum to 1500 microns.

5. Repeat Step 3.

6. Repeat Step 4 and pull the vacuum to 500 or lower microns.

7. Before shutting off the vacuum pump, close both manifold valves, then shut off the pump. Let the system sit for 15 to 30 minutes making sure the pressure does not rise above 1000 microns, which would indicate one or more of the following: a leak, trapped moisture, or trapped refrigerant.

What are Charging Methods?

The following are suggested methods for accurately charging an air conditioning system. Any of the four methods can be applied, however, the charge by weight method is the most accurate.

1. Charging by weight (weighing calculated charge).

2. System superheat (fixed orifice and cap tube).

3. Subcooling (TEV)

4. Manufacturer's charging charts.

General Charging Procedures

How Do I Charge by Weight Method?

Refrigerant is measured by weight and charged into a system at the factory or calculated by the technician after installation. Calculations are based on factory charge plus line sizes (diameter and length). An electronic charging scale should be used to weigh-in the liquid refrigerant charge, into the evacuated system. This should be done by charging the liquid directly into the liquid line service valve.

Procedure:

1. Start by placing the cylinder on the scale and connect the charging hose.

2. Zero the display of the charging scale. (The charging scale display should read zero before charging commences.)

3. Program the auto shut-off charging module to the recommended weight of charge (or monitor the scale display, stopping the flow of refrigerant when the desired charge has been added into the system).

4. If the system does not take the required charge, the remaining refrigerant should be

throttled into the suction service valve with the system operating.

5. The system should be monitored for at least 15 minutes to ensure normal operating conditions are met.

What is the Superheat Method?

The system superheat method is used to charge partially charged systems to ensure performance of a split system air conditioner or heat pump. Systems with a fixed metering device, such as a capillary tube or fixed orifice can be charged using the superheat method adjusting a charge. For a heat pump system, superheat charging can only be properly performed when the system is operating in the cooling cycle, and the outdoor temperature is above 65°F degrees.

System superheat is the heat added to the refrigerant vapor inside the evaporator all the way back to the compressor and varies as operating conditions change. For a split-system air conditioning system, the system superheat can range from 5°F to 30°F depending on indoor dry and wet bulb temperatures and the outdoor ambient temperature.

How Do You Measure System Superheat?

System superheat is a measurement of temperature increase of the saturated vapor entering the compressor. Prior to adjusting the charge, the technician must make sure the air flow is correct and the coils are clean.

Procedure:

1. Suction line temperature and pressure are taken at the outdoor unit on or near the suction service valve.

2. Saturated suction temperature is subtracted from suction line temperature. This difference in temperature is known as system superheat.

3. If superheat is low, refrigerant must be removed.

4. If superheat is high, refrigerant must be added based on the manufacturer's recommendations.

5. Most manufacturers recommend the superheat be within the range of published charging data.

What is the Subcooling Method?

If the system's metering device is a thermostatic expansion valve, the charge can be checked effectively using the subcooling method. Subcooling is based on the condenser design. Subcooling is the difference in liquid refrigerant temperature leaving the condenser and condensing saturation temperature.

Prior to adjusting the charge, the technician must make sure the air flow is correct and the coils are clean.

Procedure:

1. Liquid line temperature and pressure are taken at the outdoor unit on or near the liquid access valve.

2. Saturated liquid temperature is subtracted from liquid line temperature to determine the subcooling.

3. Recommended subcooling for each make and model of cooling or heat pump system should be supplied by the manufacturer.

4. If subcooling is high, refrigerant must be removed.

5. If subcooling is low, refrigerant must be added based on the manufacturer's recommendations.

6. Most manufacturers recommend the subcooling be within the range of published charging data.

How do I Charge Blended Refrigerants? (Liquid Charging)

Liquid charging is faster than vapor charging, but liquid charging can easily result in overcharging and can damage the compressor. It is the preferred method of charging for blended refrigerants. Liquid charging should only be performed by experienced technicians.

When adding a blended refrigerant to a partially charged system, the refrigerant must be throttled in as a liquid into the low side of the system to avoid fractionation. Safe liquid charging is accomplished using a throttling valve, or flow-reducing device, connected in series with the low side of a gauge manifold. These devices will flash the liquid into a vapor before entering the system. Frost may form on the charging hose, indicating the liquid/vapor interface. The frost line should not be allowed to reach the system. Watch the high side gauge and avoid excessive high side pressure.

Without a throttling device, the blends can separate in a partially filled refrigerant tank. When a blended refrigerant in the vapor state from a partially filled refrigerant tank is used, the refrigerant with the highest vapor pressure enters the system first.

What is Fractionation?

Fractionation is the change in the make-up of the blend because one or more of the refrigerants boils at a faster rate than the others.

How Does Fractionation Occur?

When vapor is removed from a cylinder or system containing a zeotropic blend, two things are going to happen:

1. The vapor being removed is at the wrong make-up. It will have more of the higher-pressure, higher-capacity refrigerant component.

2. The liquid that is left behind boils more of the high-pressure refrigerant out of the liquid to replace the vapor. Ultimately, the liquid changes composition because more of the higher-pressure refrigerant boils off.

In order to avoid charging the wrong refrigerant composition and fractionating the remaining blend, zeotropic blends must be removed from the cylinder as ***liquids***. To remove liquid some manufacturers suggest turning the cylinder over, so the valve is on the bottom; other manufacturers have a dip tube in the cylinder which allows only liquid refrigerant to leave the cylinder.

The worst-case scenario of fractionation occurring in a system is when approximately 1/2 of the refrigerant charge has leaked from the system.

What is Vapor Charging?

Vapor charging can only be used on systems that do not use blended refrigerants. The safest method of charging an operating system is by adding vapor refrigerant but can only be done with non-blended refrigerants (R-22, R134a, R-502 etc.).

When charging R-22 to the low side of the system, cylinder pressure must be higher than suction pressure for refrigerant to flow. Cylinder pressure can be carefully increased by warming the cylinder with hot water. Vapor charging is done with the cylinder in the upright position.

What is Psychrometrics?

Psychrometrics is the study of air and water vapor properties. For HVAC purposes standard air is established as 68°F and 29.92 inches of Hg atmospheric pressure. It occupies a space of 13.33 cubic feet per lb. and has a density of .075 lbs. per cubic foot.

A thorough knowledge of psychrometrics will help technicians to have a better understanding of moist air's properties and its effect on air conditioning.

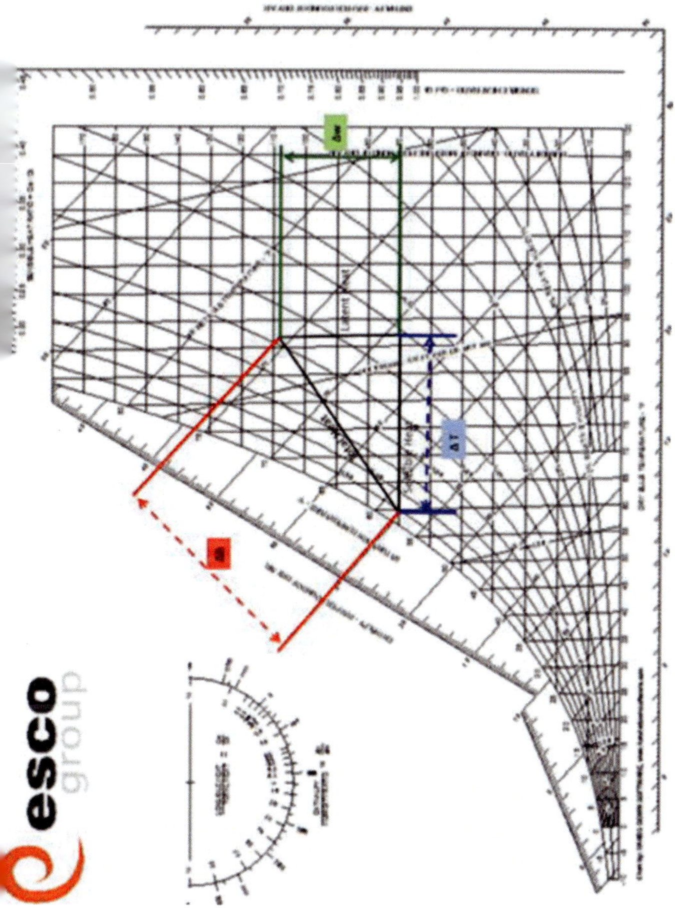

The eight psychrometric processes of air conditioning applications are implemented on the properties of air to obtain certain standards of temperature and humidity of air within the confined space. Some of the common psychrometric processes are; sensible heating or cooling of air, humidification or dehumidification of air, the mixing of indoor and outdoor air, or there may be any combination of the various processes.

Understanding these different processes is an absolute must for a technician.

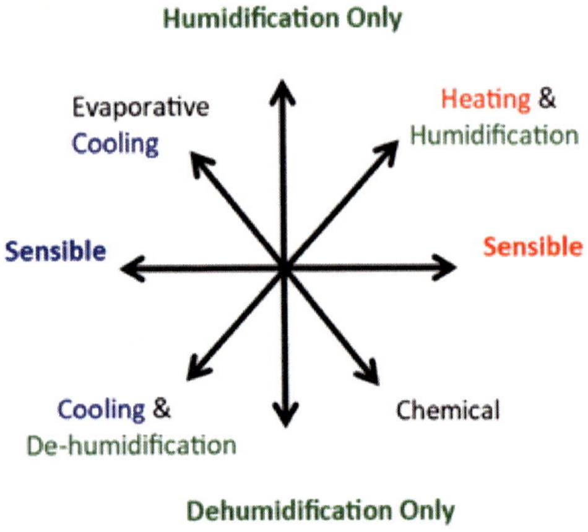

(For more in-depth information refer to *ESCO's Psychrometrics section in the System Performance series.*)

Should I Take Safety Precautions?

This manual is no substitute for safety training. It is impractical to list all safety precautions or all dangerous situations a technician may encounter during service or installation. HVACR service can be hazardous to a technician and anybody in the general vicinity. When a technician is not sure about a procedure, consult a qualified instructor or experienced service professional before proceeding. Always follow manufacturer's recommended service and safety procedures. General safety precautions can include:

- Always comply with safety data sheets (SDS).
- Always wear safety goggles when handling refrigerants or operating service equipment.
- Wear appropriate protective clothing when working with refrigerants.
- Never heat a refrigerant cylinder above 125°F.
- Refrigerant can displace oxygen causing suffocation.
- Refrigerants may cause physical injury, frostbite, or blindness.
- New refrigerants have been found to become flammable under certain conditions.
- Never fill a recovery tank above 80%.

Formulas

Square

A four-sided shape that is made up of four parallel lines with four right angles.

Area = Length x Width

A = L x W

Perimeter = 4 x any side

P = 2l + 2w

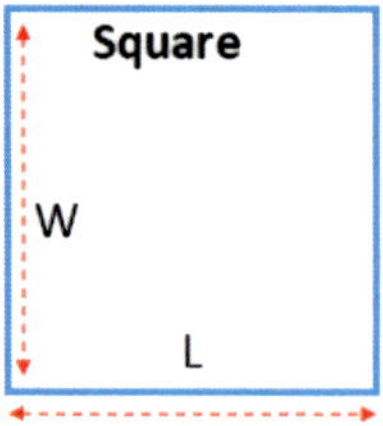

Rectangle

A four-sided shape that is made up of two pairs of parallel lines with four right angles.

Area = Length x Width

A = L x W

Perimeter = 2 x Lengths + 2 x Widths

P = 2l + 2w

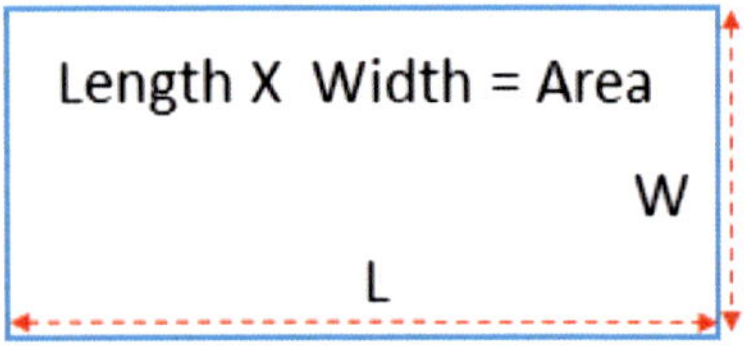

Parallelogram

A four-sided plane rectilinear figure with opposite sides parallel.

Area = Base x Height

A = BH

Perimeter = add the length of all

P = 2a + 2b

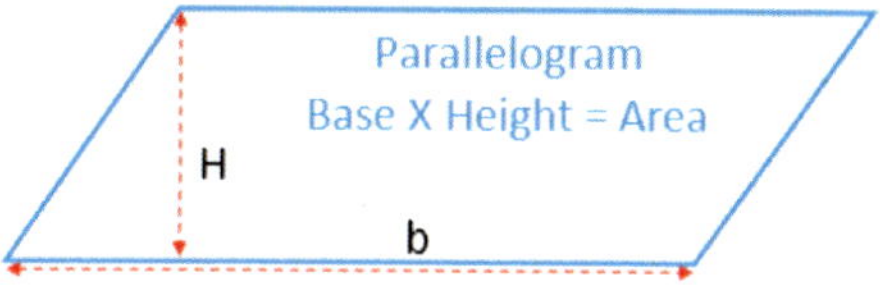

Triangle

A triangle is a 3-sided polygon every triangle has three sides and three angles.

Area = 1/2 the base x the height

A = ½ B x H

Perimeter = a + b + c

(add the length of the three sides)

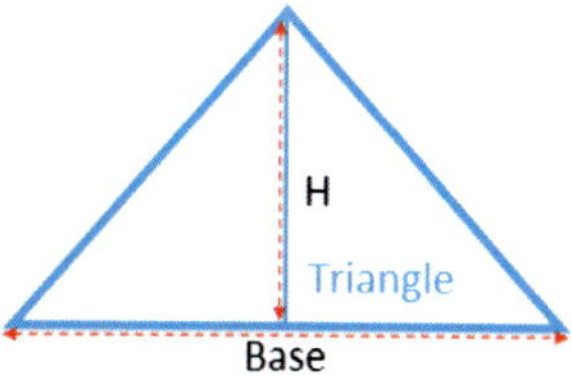

Trapezoid

A 4-sided figure with one pair of parallel sides.

Area = 1/2 the bases x the height

A = (B1 + B2 ÷ 2) x H

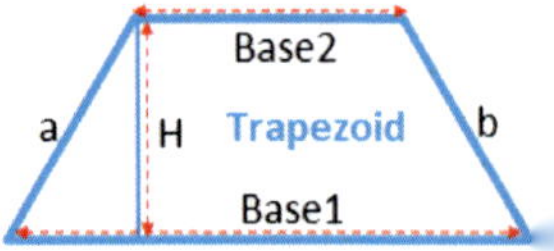

Perimeter = add lengths of all sides

P = a + b1 + b2 + c

Circle

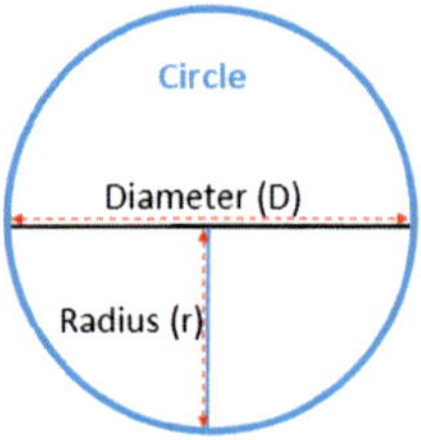

Properties of a circle

(Area $=\pi r^2$) ($\pi = 3.14$)

Center: All points on the circle are equidistant same distance from the center point.

Circumference: The circumference is the distance around the circle. ($C = d\pi = 2\pi r$)

Diameter: The distance across the circle. The length of any chord passing through the center; or twice the radius. ($D = 2r$)

Radius: The radius is the distance from the center to any point on the circle; or half the diameter. ($r = 1/2D$)

Rectangular Cube

A three-dimensional figure with three edges joined at each corner to form a vertex.

Volume = Length x Width x Height

$V = L \times W \times H$

Surface = 2lw + 2lh + 2wh

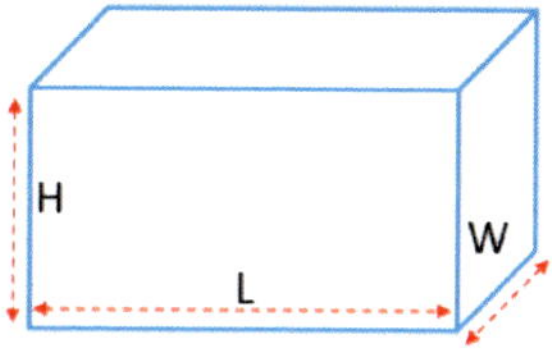

Cylinder

A three-dimensional figure with a cylindrical surface and two parallel planes.

Volume = πr^2 x height

$V = \pi r^2 h$

Surface = 2π radius x height

$S = 2\pi rh + 2\pi r^2$

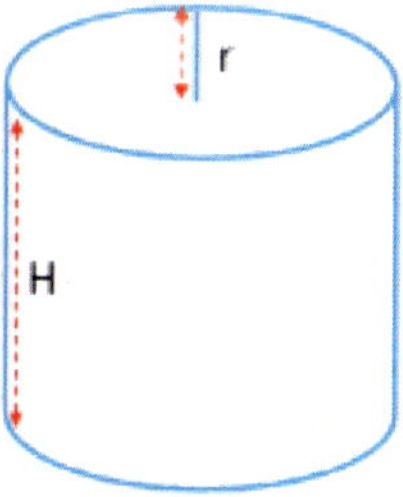

Cones

A three-dimensional figure or body having a circular base and tapering to a point at its height.

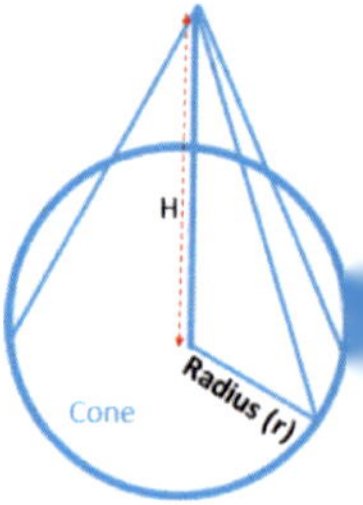

Volume = 1/3 area of the base x height

$V = \frac{1}{3}\pi r^2 h$ $V = \frac{\pi r^2 h}{3}$

Surface $S = \pi r^2 + \pi r s = \pi r^2 + \pi r \sqrt{r^2+h^2}$

Converting Measurements

- 12 Inches = 1 Foot
- 144 Square Inches = 1 Square Foot (12" x 12" = 144 Square Inches)
- 1728 Cubic Inches = 1 Cubic Foot (12" x 12" x 12" = 1 Cubic Foot)
- 3 Feet = 1 Yard
- 9 Square Feet = 1 Square Yard
- 27 Cubic Feet = 1 Cubic Yard
- 1000 Millimeters = 1 Meter
- 2.54 Centimeters = 1 Inch
- 25.4 Millimeters = 1 Inch
- 30.48 Centimeters = 1 Foot
- 1 Meter = 39.37 Inches

Converting Weight

- 1lb. = 16 ounces
- 1.5 lbs. = 24 ounces (1.5 x 16 = 24)

3 lbs. 12 ounces = (3 x 16 = 48 + 12 = 60 ounces) or (60/16 = 3.75 lbs.)

- 6 ounces = (6 / 16 = 0.375 lbs.)
- 2000 Lbs. = 1 Ton
- 1 cubic foot of water weighs 62.4 lbs.
- 62.4 lbs. ÷ 144^2 inches = 0.433
- 0.433 times the height in feet of water = psi
- Gallon of water = 8.34lbs.
- 1Lb. = 7007 Grains
- 454 grams = 1lb.
- 12 Troy ounces = 1 Troy lb.

Pressure

- Pressure = Force Divided Area
- Psi = Weight ÷ Area
- 27.7 inches WC = 1 psi
- 1Pa = 0.004 inches WC
- 0.433 x height of water in feet = Psi

Converting Temperature

- Fahrenheit = (Celsius Times 1.8) Plus 32
  (100°C x 1.8) + 32 = 212°F
- Celsius = (Fahrenheit Minus 32) Divided by 1.8
  (212°F – 32) ÷ 1.8 = 100°C
- Rankine = Fahrenheit Plus 460
  (32°F + 460) = 492° Rankine
- Kelvin = (Celsius Plus 273)
  (100°C + 273) = 373 Kelvin

Heat

The constant in Btu per hour is equal to 60 minutes x 0.075 lbs. x .24 Btu

Btuh factor = 1.08

- **Sensible heat** = 1.08 x cubic feet per minute x Temperature Difference
  1.08 x CFM x ΔT

The Btu difference between the enthalpy of 75°F water vapor and 50°F water is 1,076 Btu/lb. The enthalpy of 75°F water vapor is 1,094 Btu/lb. and the enthalpy of the 50°F (liquid) water is 18 Btu/lb. This establishes a net latent heat transfer of 1,076 Btu/lb. (1,094 Btu/lb. – 18 Btu/lb.).

(1076 Btu per lb. x 0.075lbs per cubic feet x 60 min) = 4,842
4,842 ÷ 7,000 Grains/lb. equals the 0.69 Constant

- **Latent heat** = (.69 x CFM x ΔGrains)

The constant 4.5 in the total heat formula is derived from 60 minutes per hour times the density of standard air 0.075.

(Δh = enthalpy difference)
Total Heat = (4.5 x CFM x Delta Enthalpy)

- **Total heat** = (4.5 x CFM x Δh)

Air Flow

CFM = Area (sq. ft.) x Velocity (ft. min.)

How to calculate area:

Rectangular Duct Area = L x W

Round Duct Area = $\pi D^2/4$ or πr^2

FPM = $\sqrt{VP}$ x VF and CFM = Square feet x FPM

Multiply the square root of the velocity pressure by a velocity factor of 4005 to convert to feet per minute (FPM).

Area in square feet for 8-inch round duct = $\pi r^2 \div 144$

Area = 3.14 x 42 ÷ 144
Area = 3.14 x 16 ÷ 144
Area = 50.3 ÷ 144
Area = 0.35 square feet

CFM = .35 sq. ft x 640 FPM
CFM = 224

Calculating Airflow Through a Return Air Grille

Unknown AK factor: 20" x 25" return air grille with 80 percent free area:

20 in. x 25 in. = 500 sq. in.
500 x .80 = 400 sq. in. free area
400 sq. in. ÷ 144 sq. in. per sq. ft = 2.78 sq. ft net free area or 2.8 AK factor

Multiply FPM x (AK factor or Net free area) to get CFM

SYSTEM PROBLEM	DISCHARGE PRESSURE	SUCTION PRESSURE	SUPERHEAT	SUBCOOLING	AMPERAGE
Overcharged	↑	↑	↓	↑	↑
Undercharged	↓	↓	↑	↓	↓
Liquid Re-striction	↓	↓	↑	↑	↓
Low Evap Air-	↓	↓	↓	↓	↓
Dirty Condenser	↑	↑	↓	↓	↑
Low Outdoor Ambient	↓	↓	↑	↑	↓
Insufficient Compressor	↓	↑	↑	↑	↓
TXV Bulb Charge Lost	↓	↓	↑	↑	↓
TXV Bulb Loose	↓	↑	↓	↓	↓
High Heat Load	↑	↑	↑	LOW-NORMAL	↑
Restricted Me-tering Device	↑	↓	↑	↑	—
R-22: SUCTION – 70 PSIG, DISCHARGE – 250 PSIG *R-410A: SUCTION – 135 PSIG, DISCHARGE – 320 PSIG*			*SUPERHEAT: 15°F* *SUBCOOLING: 10°F*		

HVAC Excellence APPROVED™

ISBN: 1-930044-03-8

9781930044036